never forget numbers & dates!

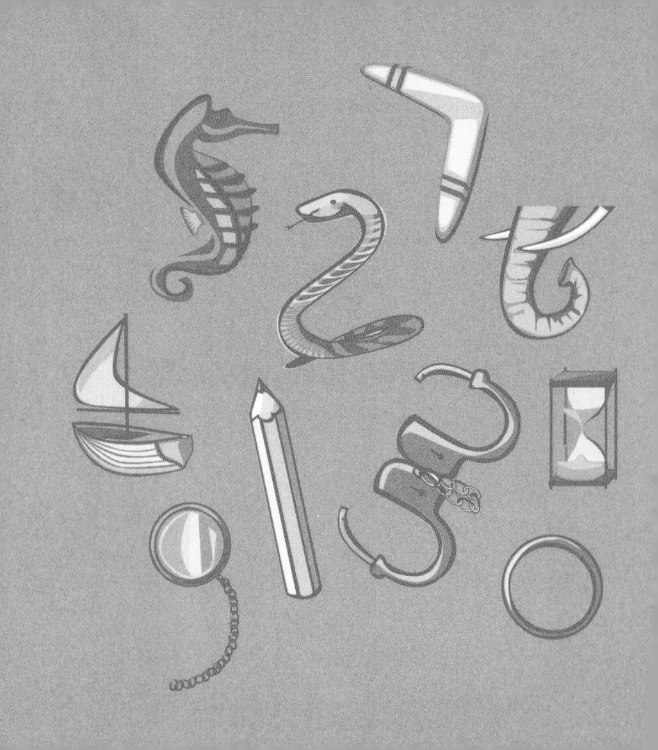

DOMINIC O'BRIEN

never forget numbers & dates!

To all who regard memory
as an art and a science.

Never Forget!
Numbers and Dates

Dominic O'Brien

First published in the United Kingdom and Ireland in 2002 by
Duncan Baird Publishers Ltd
Sixth Floor
Castle House
75–76 Wells Street
London W1T 3QH

Conceived, created and designed by Duncan Baird Publishers

British Library Cataloguing-in-Publication Data:
A CIP record for this book is available from the British Library.

ISBN: 1-903296-82-X

10 9 8 7 6 5 4

Typeset in Helvetica Condensed
Printed and bound in Thailand
by Imago

CONTENTS

INTRODUCTION

Over the past few decades we have been concentrating on keeping our bodies in shape by exercising and going to the gym. Now, increasingly, we are turning toward training our minds.

As World Memory Champion I need to keep my mind in tune with regular mental exercises using numbers. One of these involves listening to a long number randomly generated and spoken to me by my computer. I typically memorize a 200-digit number spoken at the rate of 1 digit per second, which I hear only once. Yet 14 years ago, before training my memory, I could recall only 7 or 8 digits – about the length of a phone number – which is the average human memory span for numbers. ›

In this book I will show you how to breathe life into what seems like meaningless data. Using techniques such as the Number–Rhyme System, the Number–Shape System, and my own DOMINIC System, which has helped me win the World Memory Championships eight times, you will find the language of numbers easy to grasp, fun to learn and of great practical benefit.

We live, increasingly, in a world dominated by numbers. I hope that the techniques presented in this book will make you look at numbers in a more positive and memorable light.

Dominic O'Brien

SYMBOLS USED IN THIS BOOK

MEMORY TECHNIQUE

MEMORY WISDOM

MEMORY IN ACTION

THE IMPORTANCE OF NUMBERS AND DATES

We organize our lives around numbers – important dates ranging from work deadlines to family birthdays, and times for appointments and social events. And as using technology such as computers, mobile phones and cash machines has become part of everyday life, we find ❯

ourselves having to remember more and more security codes, phone numbers and PINs.

So why do we have such trouble remembering them? First, our brains have not evolved a specific ability to deal with numbers, and second, most of us (except mathematicians!) find numbers uninspiring. The challenge, then, is to make numbers interesting and hence more memorable.

THE GREEK CONNECTION

The ancient Greeks devised a technique that has become widely used to remember numbers and dates – it's called mnemonics. Often this method involves the use of words to make numbers more meaningful. For example, you might try to memorize the PIN number 1425 by constructing a phrase consisting of words that have the same number of letters ›

5

as each digit, such as "I (1) save (4) my (2) money (5)." Or you might memorize your office entry code, 3258, as "How (3) to (2) enter (5) building (8)." Or, again you might memorize your house alarm code, 5352, as "No-one (5) can (3) break (5) in (2)." Note that it helps if your phrase has a connection with whatever the number is used for, but this is not essential.

MNEMONIC RHYMES

You might have used mnemonics at school, perhaps in the form of rhymes that help you recall dates, for example: "Columbus sailed the ocean blue in fourteen hundred and ninety-two", and "The Spanish Armada met its fate in fifteen hundred and eighty-eight." You could try creating your own to help you remember significant years. >

1492
1588
1983

LET'S SAY, FOR EXAMPLE, THAT YOU ARE ALWAYS FORGETTING HOW OLD A FRIEND'S SON IS, OR HOW LONG YOUR COUSIN HAS BEEN MARRIED. YOU COULD MAKE UP RHYMES LIKE THESE: "IN NINETEEN HUNDRED AND EIGHTY ONE, PETER AND JANET HAD JAMES, THEIR SON," OR "JASON MARRIED HIS GIRLFRIEND KATE IN NINETEEN HUNDRED AND NINETY EIGHT," OR STILL AGAIN, "IN NINETEEN HUNDRED AND FIFTY NINE WAS BORN THAT WONDERFUL MOTHER OF MINE."

WITH A LITTLE INGENUITY AND PERSEVERANCE YOU COULD SOON BUILD UP A GROUP OF YOUR OWN RHYMES CONTAINING THE DATES OF ALL YOUR FAMILY AND FRIENDS' BIRTH YEARS AND IMPORTANT ANNIVERSARIES, SO THAT YOU NEVER FORGET ANY OF THEM AGAIN.

GINGKO BILOBA

Can dietary supplements improve your memory? Research has suggested that taking extract of gingko biloba, either in tablet or liquid form, or as tea, can help memory function by increasing the blood flow to the brain. I always take gingko biloba as part of my preparations before a memory competition.

THE NUMBER–RHYME SYSTEM

Rhymes also feature in the next technique – the Number–Rhyme System. You pick a number and then you think of a word that rhymes with it. For example, "one" rhymes with "bun", "two" with "shoe", and so on. The word you select then becomes your key image for that number, and always represents that number in memorizations using this system. >

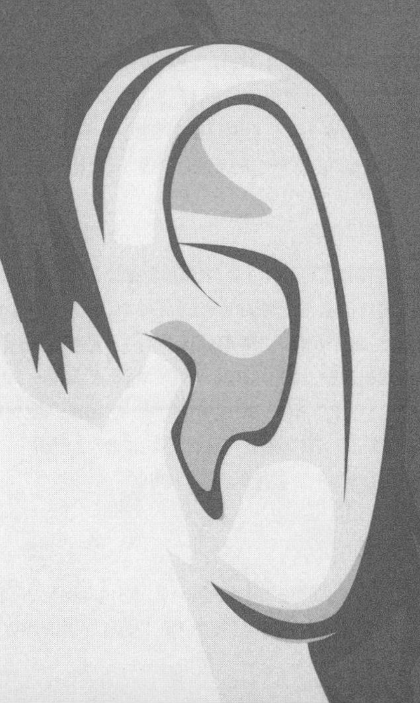

RESEARCH HAS SHOWN THAT THE NUMBER–RHYME SYSTEM WORKS BEST WHEN YOU CREATE YOUR OWN ASSOCIATIONS, SO I WOULD NOW LIKE YOU TO SPEND A FEW MINITES THINKING UP NUMBER–RHYMES FOR THE DIGITS 0 TO 9.

IN MY EXPERIENCE NOUNS (WORDS DESCRIBING THINGS OR PLACES) TEND TO WORK BETTER THAN VERBS (WORDS DESCRIBING ACTIONS) OR ADJECTIVES (WORDS THAT DESCRIBE NOUNS), BECAUSE THEY ARE EASIER TO IMAGINE. HOWEVER, THERE ARE NO RIGHT OR WRONG CHOICES. THE IMPORTANT THING IS THAT EACH OF YOUR NUMBER–RHYMES WORKS FOR YOU.

HOW DID YOU GET ON? OPPOSITE, IS A LIST OF SOME POPULAR EXAMPLES TO COMPARE WITH YOUR OWN.

0 is a hero

1 is a bun, a nun or the sun

2 is a shoe or some glue

3 is a tree, a bee or a knee

4 is a door, a saw or a paw

5 is a hive or a dive

6 is some sticks or some bricks

7 is heaven

8 is a gate, a date or a weight

9 is some wine, a line or a sign

Let's now try an example. Imagine you bump into an old friend in the street and he gives you his phone number so that you can arrange to meet up. The number is 419 3027.

To remember the number using number–rhyme you visualize opening your front door (4), and going outside to eat a bun (1) and drink a glass of wine (9), sitting under a tree (3) in your garden. Suddenly your hero (0) (anyone you admire) rushes up carrying a shoe (2) – perhaps a glass slipper as in the story of Cinderella, or a golden football shoe if you are sports fan – and this makes you so happy that you feel "in seventh heaven"(7).

Try creating your own story using your own telephone number, as practice.

1074
392

MEMORY FOODS

It is vital to eat a healthy diet if we wish to keep our memory in full working order. Foods rich in vitamins A, C and E, such as oranges and red peppers, have been shown to aid memory recall. Oily fish, such as salmon, are another rich source of "brain food". Try to eat oily fish at least twice a week.

THE NUMBER–SHAPE SYSTEM

Another good way to make numbers memorable is through the Number–Shape system, in which we use our creativity to turn numbers into objects that resemble the figure's written shape.

Let's now take an in-depth look at the shapes that are often associated with the numbers 0 to 9. >

The number zero can be shown as a ball, a wedding ring, or the sun.

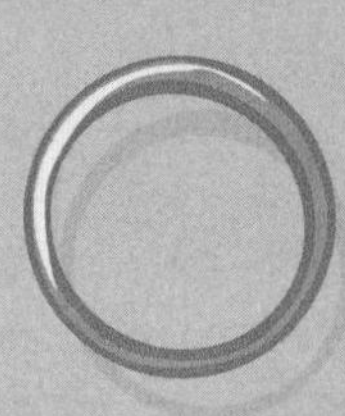

0

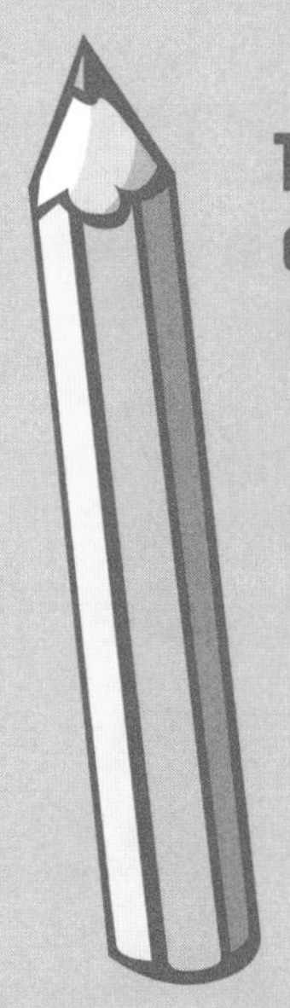

1

The number one can be represented by a pencil or a candle, but a matchstick, a rocket or a knife would work equally well. ›

The number two can be shown as an elegant swan, or as

a snake rearing as if to strike.

2

The number three can be depicted as the profile of lips puckered in readiness for a kiss, or by handcuffs. ›

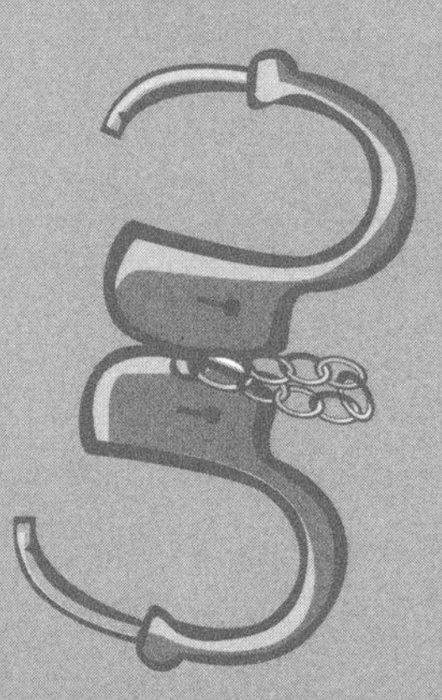

The number four can be shown as the unfurled sail on a boat, or a flag, fluttering in the wind.

The number five can be represented by a graceful sea-horse or a fish hook. >

5

The number six can be shown as an elephant's trunk,

6

a golf club or a croquet mallet.

The number seven can be represented by a boomerang, the edge of a cliff or a diving board. >

The number eight can be shown as a snowman,

or as an egg-timer or hourglass.

And finally, the number nine can be represented as a monocle or a balloon on a string. ›

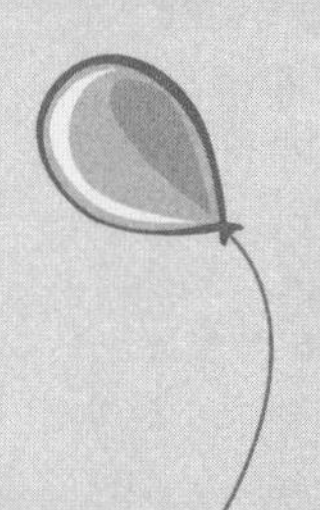

Of course, if you prefer, you can think up your own number–shapes. The important thing is to choose only one shape for each number – you are aiming to build up a straightforward visual alphabet of numbers and shapes that have strong or automatic associations for you.

To practise your skill at number–shape recognition, try the following exercise. All you do is read through the rows of numbers (opposite), from left to right as if you were reading a book, and convert each number into the number–shape you have chosen to represent it. Don't try to memorize the sequence at first, simply use the numbers as an exercise in making the associations. Start slowly – as you go along

YOU WILL PROBABLY FIND THAT YOU GET FASTER. THEN, WHEN YOU FEEL CONFIDENT, TRY MEMORIZING THE FIRST ROW BY CREATING A STORY USING THE NUMBER–SHAPES (JUST AS YOU DID WITH THE NUMBER–RHYMES). AS YOU GET MORE PROFICIENT DO THE SAME WITH MORE ROWS UNTIL YOU HAVE MEMORIZED ALL 30 NUMBERS.

7	1	3	5	0
8	2	4	9	1
6	4	1	5	7
0	2	9	3	8
5	6	0	7	1
4	9	4	2	3

›

LET'S LOOK AT AN EXAMPLE. HERE, BY CREATING A VISUAL SYMBOL FOR EACH NUMBER, WE ARE GOING TO TURN AN INSTANTLY FORGETTABLE LIST OF DIGITS INTO IMAGES THAT CAN BE LINKED TOGETHER INTO A VIVID, MEMORABLE SCENE.

DRAWING UPON THE NUMBER–SHAPES THAT WE HAVE JUST LEARNED, LET'S MEMORIZE THE PIN NUMBER 8190. USING A SNOWMAN AS 8, A PENCIL AS 1, A BALLOON WITH STRING ATTACHED AS 9, AND THE SUN AS 0, THE SEQUENCE 8, 1, 9, 0 BECOMES A LARGE, FRIENDLY-LOOKING SNOWMAN, WHO HAS A PENCIL AS HIS NOSE AND HOLDS A BALLOON ON A STRING AS HE MELTS IN THE SUN. SPEND A FEW MINUTES NOW VISUALIZING THE SCENE IN AS MUCH DETAIL AS POSSIBLE.
NOW COVER THIS PAGE.

What was the number? Try this technique to memorize your own PINs and codes.

FIRST IS BEST

When making associations, always stick with the first image that comes into your mind (no matter how absurd this mental picture may seem to you on later reflection). Your brain's initial reaction is intuitive – the image that first comes to mind is the one that you are most likely to recall automatically.

MEMORY AND SLEEP

Getting enough rest is vital if we wish to have a good memory – while we are asleep our brain consolidates the events of the day. Try a pre-sleep "breath meditation" to relax you into slumber. Close your eyes and take deep, slow breaths for 5 minutes. Try to make each complete breath last for a slow count of 10.

"CHUNKING"

So far we have learned how to memorize lists of digits, by making a new association for each number. Now we shall learn a quick method of memorizing long sequences of numbers – "chunking". As the word implies, the technique involves dividing long numbers into small groups that are easier to assimilate than long strings of digits. ›

1066

1733

1819

LET'S TAKE A LOOK AT AN EXAMPLE. IMAGINE THAT YOU ARE TRYING TO MEMORIZE THE FOLLOWING NUMBER: 117578338124 – PERHAPS IT IS YOUR BANK ACCOUNT NUMBER. AT FIRST GLANCE YOU MIGHT WONDER HOW YOU COULD EVER REMEMBER SUCH A LONG, MEANINGLESS STRING OF DIGITS. HOWEVER, IF YOU GROUP SOME OF THE NUMBERS TOGETHER, THEY IMMEDIATELY BECOME

MORE MEANINGFUL, AND THEREFORE MORE MEMORABLE. FOR EXAMPLE, TAKING THE FIRST TWO DIGITS "11", YOU MIGHT THINK OF THE BINGO CALL "LEGS 11"; THEN, THE NEXT THREE NUMERALS "757" MIGHT REMIND YOU OF THE TYPE OF AIRPLANE THAT TOOK YOU ON HOLIDAY; "8338" IS EASY TO REMEMBER BECAUSE IT IS A PALINDROME – IT READS THE SAME FORWARD AND BACKWARD – AND "124" MIGHT BE THE NUMBER OF THE BUS YOU OFTEN CATCH INTO TOWN.

NOW COMBINE THESE IMAGES IN A SHORT STORY, WHICH WILL FIX THE SEQUENCE OF THE NUMBERS IN YOUR MIND.

EVOCATIVE SCENTS

By burning scented candles or using certain aromatherapy oils we can increase our powers of recall. Try this bathtime memory enhancer. Sprinkle a few drops of lemon, basil or sandalwood essential oil into your bathwater. Breathe slowly and relax. Then, close your eyes and revisit your most pleasant memories.

NUMBER ASSOCIATIONS

There are certain numbers that immediately conjure up particular associations. For example, if you asked me what I associate with the number 3, I'd say "the three Wise Men". Or if you asked me what I associate with the number 7, I'd reply: "the seven dwarfs" (Snow White's friends). What associations do you immediately make with certain numbers? ›

To find out, try the following exercise. Take a piece of paper and write the numbers 0 to 9 in a column down the left-hand side of the page. Now focus on each number in turn and write down the first association that springs to mind. Don't sit wracking your brain. If you can't think of anything straight away, leave that number and move on to the next. Don't worry about any gaps that there may be in your list because you are only trying to bring to light any links you have already (unconsciously) made, which will help you improve your memorization of numbers.

When you have finished, make a mental note of the associations you have discovered – they will be effective memory tools precisely because they are instinctive links.

1
2
3
4
5

THE THREE KEYS TO MEMORY

Now we're going to build on what you've learned so far. You have already used your *imagination* to make a strong link between each number and its *association*. Next, we'll link the associations to a *location* to make their order infallibly memorable. And in so doing, we'll use the three keys for memorization: *imagination*, *association* and *location*. >

THE JOURNEY METHOD

The ancient Greeks discovered that by attaching mental pictures of the things they needed to remember to a sequence of places, such as seats around a table, they could give unrelated objects a logical order. When applied to numbers, this method makes digits easier to recall in a particular order, as you will discover in the Journey Method. >

In this technique you conjure up a clear mental picture of a real journey you know well, such as your daily walk to the train station or the route you drive your children to school. Then, you mentally place the numbers you wish to memorize along that journey, at fixed stages. When you need to recall the numbers all you have to do is simply mentally retrace your steps.

CLOSE YOUR EYES AND IMAGINE YOURSELF SETTING OFF FROM YOUR HOME ON A JOURNEY YOU KNOW REALLY WELL. VISUALIZE YOURSELF GOING THROUGH THE DOOR, DOWN THE PATH, THROUGH THE GATE. AS YOU WALK OR DRIVE ALONG, WHAT ARE THE LANDMARKS THAT CATCH YOUR EYE? FOR EXAMPLE, DO YOU ALWAYS PASS THE POLICE STATION OR THE HOSPITAL? AS YOU RECALL THE JOURNEY IDENTIFY TEN MEMORABLE, FIXED STAGES THAT YOU ALWAYS PASS ALONG THE ROUTE, AND IMAGINE THEM AS VIVIDLY AS YOU CAN, IN THE SAME ORDER THAT YOU COME TO THEM. >

ONCE YOU HAVE IDENTIFIED ALL THE STAGES OF YOUR MENTAL JOURNEY, YOU CAN START TO USE THEM TO "STORE" YOUR NUMBERS. YOU CAN USE ANY OF THE TECHNIQUES WE HAVE COVERED TO DO THIS, SO FEEL FREE TO EXPERIMENT AND DISCOVER WHAT WORKS BEST FOR YOU. I FIND THAT ONE OF THE MOST EFFECTIVE METHODS IS TO COMBINE THE JOURNEY METHOD WITH THE NUMBER–SHAPE SYSTEM.

WORKING THROUGH EACH STAGE OF YOUR JOURNEY IN TURN, YOU PLACE THE NUMBERS THAT YOU NEED TO REMEMBER ALONG THE ROUTE. LET YOUR IMAGINATION RUN RIOT TO CREATE IMAGES THAT ARE AS DRAMATIC AND DETAILED AS YOU CAN TO CONNECT THE DIGITS WITH THEIR RESPECTIVE STAGES – THE MORE BIZARRE THE IMAGES, THE MORE MEMORABLE THEY ARE LIKELY TO BE.

LET'S LOOK AT AN EXAMPLE. WE'LL ASSUME THAT YOU ARE MEMORIZING YOUR BANK ACCOUNT NUMBER: 0178256342, USING LANDMARKS ON YOUR DAILY WALK TO THE TRAIN STATION.

THE STAGES ON YOUR JOURNEY ARE: THE PARK, THE FIRE STATION, THE DINER, THE BUS STOP, THE HAT STORE, THE JEWELRY STORE, THE TRAVEL AGENCY, THE POLICE STATION, THE PARKING LOT AND THE FLOWER STALL. >

THE FIRST NUMBER IS 0, WHICH YOU PLACE AT THE FIRST LANDMARK – THE PARK. USING ONE OF THE ASSOCIATIONS YOU LEARNED IN THE NUMBER–SHAPE METHOD, YOU COULD IMAGINE A HUGE BALL SITTING IN THE MIDDLE OF THE GRASS.

OR, IF 0 AUTOMATICALLY MAKES YOU THINK OF A WHEEL, VISUALIZE A FAIRGROUND FERRIS WHEEL WITH LOTS OF NOISY CHILDREN ON BOARD.

NEXT IS 1, WHICH YOU NEED TO LINK WITH THE FIRE STATION, SO YOU COULD CONJURE UP IN YOUR MIND'S EYE A GIANT PENCIL BLOCKING THE FIRE STATION DOORS, WITH THE FIREMEN TRYING TO HAUL IT AWAY WITH ROPES. OR, IF THIS NUMBER MAKES YOU THINK OF, SAY, A STREET LAMP YOU COULD IMAGINE A LAMP OUTSIDE THE FIRE STATION CATCHING FIRE. >

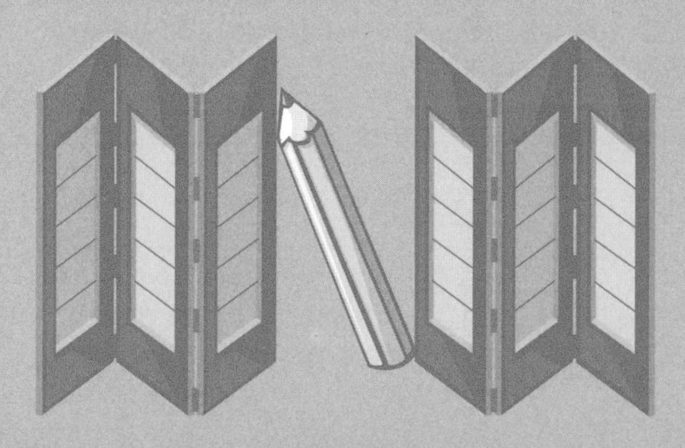

YOU NOW MOVE ON TO THE THIRD NUMBER, 7, WHICH YOU ARE GOING TO LINK WITH THE DINER. AS ITS SHAPE REMINDS YOU OF A DIVING-BOARD, YOU COULD VISUALIZE THE "SOUP-DIVING CHAMPIONSHIPS" TAKING PLACE IN THE DINER, WITH CONTESTANTS DIVING FROM THE HIGH BOARD INTO A POOL OF LUKEWARM SOUP!

OR IF 7 IMMEDIATELY SUGGESTS TO YOU THE SEVEN DEADLY SINS, YOU COULD IMAGINE A GLUTTON EATING HIS OR HER WAY THROUGH A VAST ARRAY OF FOOD AT A TABLE IN THE DINER.

THE FOURTH NUMBER IS 8 AND THE FOURTH STAGE IS THE BUS STOP. IF YOU FIRST THINK OF AN IMAGE USING THE NUMBER–SHAPE SYSTEM, YOU COULD IMAGINE A GIGANTIC EGG-TIMER SHOWING THE COUNTDOWN TO THE ARRIVAL OF THE NEXT BUS.

OR, IF THE NUMBER EIGHT CONJURES UP A MENTAL PICTURE OF, SAY, AN OCTOPUS, YOU COULD VISUALIZE THE SEA CREATURE STANDING AT THE BUS STOP WAVING ITS EIGHT TENTACLES TO FLAG DOWN A BUS. >

NEXT COMES 2, WHICH YOU NEED TO LINK TO THE HAT STORE. IF THE SHAPE OF THIS NUMBER SUGGESTS A SWAN, YOU COULD IMAGINE THE SALES ASSISTANT WEARING A LARGE WHITE HAT FEATURING A GLIDING SWAN. OR, SAY, THE NUMBER 2 MAKES YOU THINK OF TWINS, YOU COULD VISUALIZE TWIN SALES ASSISTANTS WORKING IN THE HAT STORE.

THE SIXTH NUMBER IS 5, WHICH YOU PLACE AT THE JEWELRY STORE. USING NUMBER–SHAPE IMAGES, YOU COULD VISUALIZE A SEAHORSE PENDANT MADE OF BRILLIANT DIAMONDS DISPLAYED IN THE STORE WINDOW. OR, IF THE FIRST THING THAT COMES TO MIND IS, SAY, THE PHRASE "FIVE GOLD RINGS" FROM THE CHRISTMAS CAROL "THE TWELVE DAYS OF CHRISTMAS", YOU COULD IMAGINE FIVE GIANT, INTERLINKING GOLD RINGS (IN THE MANNER OF THE OLYMPIC GAMES EMBLEM) IN THE WINDOW. >

You move on now to the number 6, which you are going to link with the travel agency. If the shape of this number reminds you of an elephant's trunk, you could picture an elephant standing in the window, advertising safari holidays in Africa.

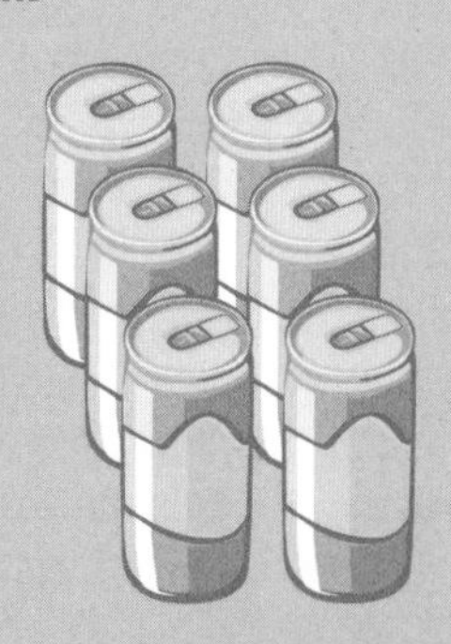

Or if 6 makes you think of, say, a six-pack of beer, you could imagine the window dressed like a German beer hall, stacked high with six-packs.

THE NUMBER 3 IS NEXT, AND YOU NEED TO ASSOCIATE IT WITH THE POLICE STATION. YOU COULD IMAGINE A POLICEMAN OUTSIDE HOLDING A PAIR OF HANDCUFFS.

OR IF THE FIRST THING THAT COMES TO MIND IS, SAY, A TRIANGLE, YOU CAN IMAGINE A LARGE TRIANGULAR LIGHT ABOVE THE ENTRANCE TO THE POLICE STATION. >

THE NINTH NUMBER IS 4, WHICH YOU PLACE IN THE CAR PARK. USING AN IMAGE FROM THE NUMBER-SHAPE SYSTEM, YOU COULD CONJURE UP A LARGE SAILBOAT THERE.

OR IF 4 MAKES YOU THINK OF THE FOUR POINTS OF A COMPASS, YOU COULD IMAGINE A HUGE SIGN IN THE MIDDLE OF THE PARKING LOT, INDICATING IN WHICH DIRECTIONS LIE NORTH, SOUTH, EAST AND WEST.

LASTLY WE HAVE THE NUMBER 2 AGAIN, WHICH YOU ARE GOING TO LINK TO THE FLOWER STALL.

IF YOU ARE USING THE SHAPE OF A SWAN TO REPRESENT THIS NUMBER, YOU COULD VISUALIZE A SWAN MADE FROM FLOWERS AS THE CENTREPIECE ON THE STALL. OR IF YOU IMMEDIATELY THOUGHT OF TWINS AGAIN, YOU COULD IMAGINE TWIN FLOWER-SELLERS. >

Now, you have linked each digit to a stage on the journey. When you need to recall your bank account number, you just mentally walk the journey. As you reach each stage, the associations (and through them, the digits) should spring to mind immediately. But don't worry if at first it takes you a while to recall them – all you need to do is practise. >

3

5

9

7

6

4

1

USE YOUR SENSES

One way of making your mental images as vivid and memorable as possible is to involve all your senses in your associations. For example, "smell" the aromas of the food wafting out of the diner, "hear" the elephant trumpeting in the window of the travel agency, "feel" the wind blowing across the parking lot, and so on.

BUILD UP A STOCK OF JOURNEYS

Once you have mastered the Journey Method, you will be able to remember long sequences of digits. As it is easier to remember several short journeys rather than one long journey, I suggest you start by thinking up several journeys of ten stages each. You can then expand them to, say, 25 stages, when you get more confident.

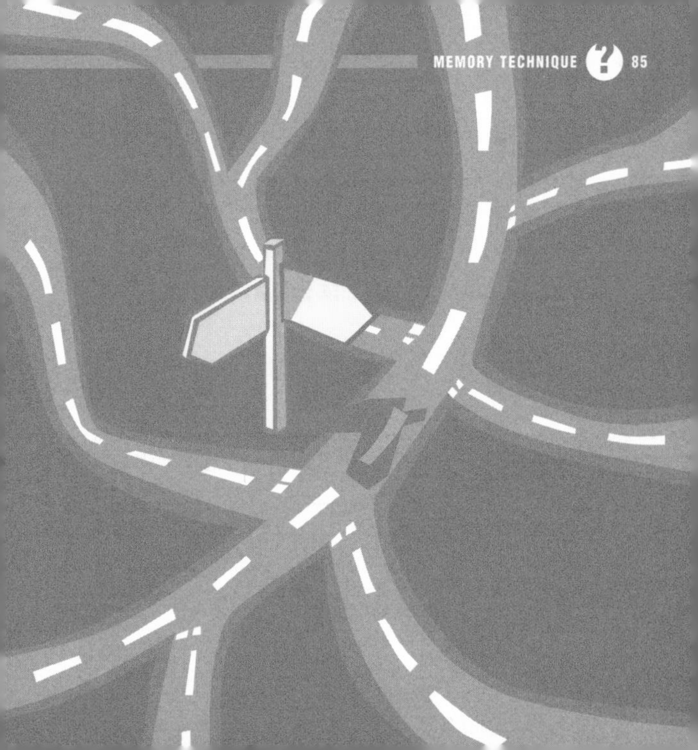

IT'S TIME TO TRY OUT YOUR NEWLY-AQUIRED SKILLS FOR USING THE JOURNEY METHOD. TRY MEMORIZING THE FOLLOWING LIST OF NUMBERS, 7316458902, USING YOUR OWN TEN-STAGE MENTAL JOURNEY (YOU CAN USE THE ONE YOU DEVISED ON P.69 IF YOU LIKE). ALLOW 10 MINUTES FOR MEMORIZING AND 5 MINUTES FOR RECALL.

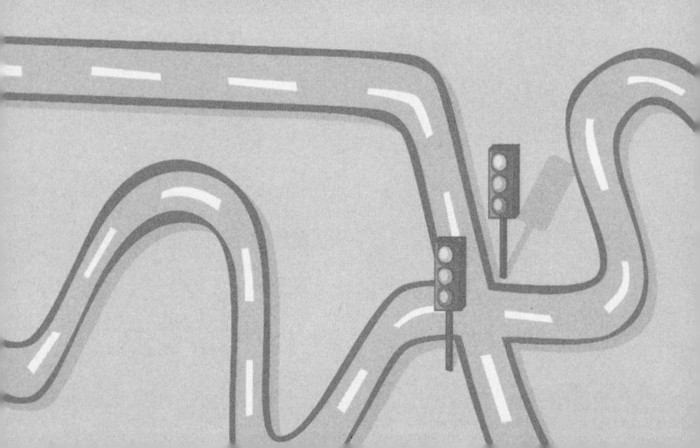

HOW DID YOU GET ON? IF YOU FOUND IT DIFFICULT, DON'T WORRY. JUST KEEP PRACTISING – YOU WILL BE SURPRISED AT HOW SOON YOU IMPROVE. ONCE YOU FIND IT RELATIVELY EASY TO MEMORIZE A TEN-DIGIT NUMBER, TRY TO MEMORIZE A NUMBER OF 20 DIGITS' LENGTH, BY CREATING A SECOND 10-STAGE MENTAL JOURNEY AND RUNNING THE TWO JOURNEYS TOGETHER.

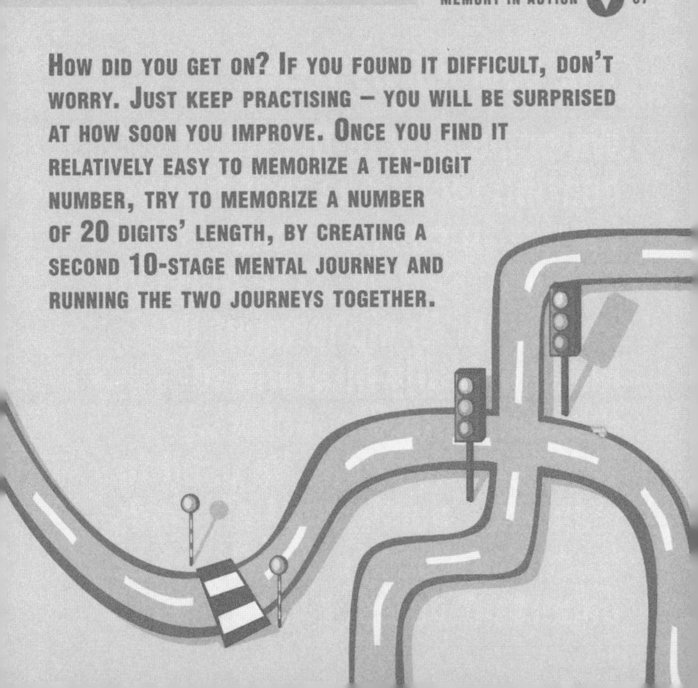

THE DOMINIC SYSTEM

We come now to the world-beating technique that I devised for use in competitions – the DOMINIC System. (The acronym stands for Decipherment Of Mnemonically Interpreted Numbers Into Characters.) This technique can be used to memorize numbers of all kinds, from statistics to equations to important dates. I have even ›

67

used it to memorize the Periodic Table and the entire calendar of the 21st century!

I wanted to create a method that enabled me to translate

numbers into images instantly. After much experimentation I decided that the best way to do this was to associate the numbers with familiar people, because people are interesting, and hence memorable.

Here's how I did it. First, I formulated a number–alphabet in which the numbers 0 to 9 are each represented by a letter, as set out on the following page. ›

0 is O (because they look similar)
1 is A
2 is B
3 is C
4 is D
5 is E
(because A to E are the first five letters of the alphabet)
6 is S (because of the "s" sound)
7 is G (because G is the 7th letter of the alphabet)
8 is H (because H is the 8th letter of the alphabet)
9 is N (because the word "nine" contains two "n"s)

Next, I substituted these letters for all the numbers from 0 to 99. With the single digits 0 to 9, I put a 0 in front of each number to turn it into a pair, like this: 00, 01, 02, and so on. The result is a series of 100 pairs of letters. I think of these pairs as initials of personalities or people known to me whose names start with these initials. I found that ❯

I didn't need to visualize each person's face clearly (although it was helpful if I could), but rather that I needed to associate each person with their own unique, characteristic action or particular prop. For example, the number 11 converts to AA, which is represented by Andre Agassi in my list. An obvious choice of prop for Andre Agassi, as a tennis player, is a tennis racket.

Sometimes it might be difficult to think of someone with the relevant initials. The number 07, which converts to OG, is one such stumbling block for me. In this case I thought of a particular profession instead: OG became Organ Grinder.

Eventually, all the numbers became personalities or people in my own special technique – the DOMINIC system. >

LET'S NOW LOOK AT THE DOMINIC SYSTEM IN ACTION. SAY YOU WERE TRYING TO REMEMBER THE NUMBER 7268503491 (10 DIGITS). AS THIS IS A LONG NUMBER, THE FIRST THING YOU NEED TO DO IS BREAK IT DOWN INTO PAIRS OF DIGITS (72, 68, 50 AND SO ON) WHICH THEN BECOME PAIRS OF INITIALS: GB (72), SH (68), EO (50), CD (34) AND NA (91).

LET'S ASSUME THAT THESE INITIALS STAND FOR GEORGE BUSH, SADDAM HUSSEIN, EUGENE O'NEILL, CELINE DION AND NEIL ARMSTRONG. YOU NOW HAVE THE CHARACTERS, ALL YOU HAVE TO DO IS FIX THEM IN YOUR MIND. AND THE BEST WAY TO DO THIS IS TO CHOOSE ONE OF YOUR MENTAL JOURNEYS AND LINK EACH CHARACTER TO A STAGE ALONG IT, FOLLOWING THE ORDER IN WHICH THE NUMBERS THAT ›

K-142

YOU ARE TRYING TO REMEMBER APPEAR. LET'S SAY YOU CHOOSE A JOURNEY AROUND YOUR HOME. YOU FIRST ENTER THE FRONT ROOM WHERE YOU SEE GEORGE BUSH SITTING ON THE SOFA WATCHING TV, FLANKED BY TWO PRESIDENTIAL BODYGUARDS. THEN YOU GO INTO THE DINING ROOM AND FIND SADDAM HUSSEIN DIRECTING A MILITARY CAMPAIGN WITH TOY TANKS ON THE TABLE.

MOVING INTO THE KITCHEN, YOU ENCOUNTER EUGENE O'NEILL TAKING A LARGE BAG OF ICE CUBES OUT OF THE FRIDGE. NEXT YOU GO UPSTAIRS INTO THE BATHROOM. YOU FIND CELINE DION SINGING AT THE TOP OF HER VOICE AS SHE RELAXES IN THE BATH AMID A SEA OF BUBBLES. FINALLY YOU VISIT YOUR SON'S BEDROOM WHERE NEIL ARMSTRONG, WEARING A SPACE SUIT, IS BOUNCING ON THE BED IN SLOW MOTION, AS IF HE IS ON THE MOON. >

As well as being very handy for memorizing numbers, the DOMINIC system is also useful for memorizing dates.

Let me show you how. This time, let's imagine that you wish to memorize the birthdays of your relatives Aunt Wendy, Uncle John, Great-aunt Heather, cousin Freddie, cousin Damian and cousin Laura. Listing their birthdays by date (the day first, followed by the month) you have: 27/02, 21/03, 13/07, 12/08, 09/10, 23/11.

You now turn these dates into initials and think up personalities whose names start with those initials. Let's say you have BG/OB (Bob Geldof and Otto von Bismarck), BA/OC (Bryan Adams and Ornette Coleman), AC/OG (Agatha Christie and an Organ Grinder), AB/OH (Annette >

Bening and Oliver Hardy), ON/AO (Oliver North and Annie Oakley), and BC/AA (Bill Clinton and Andre Agassi).

Note that here, where it would at first appear that you need to link two personalities with each location, you can instead "condense" each image by imagining the first person with the second person's prop, or carrying out the second person's action – you will see this demonstrated in the examples on pp.104–5.

First you need to link the personalities you have chosen with your relatives as vividly and with as much attention to detail as you can. One of the best ways to do this is to mentally place them in scenes that you visualize as being in- or outside your family members' homes. >

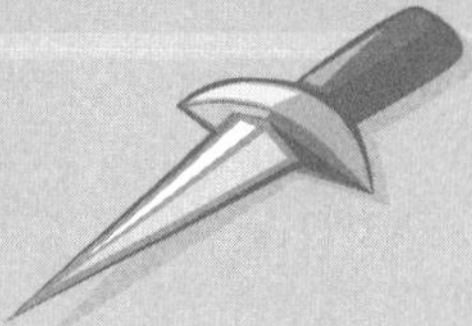

LET'S LOOK AT SOME EXAMPLES. AT AUNT WENDY'S HOUSE YOU IMAGINE BOB GELDOF PLAYING WITH TOY SOLDIERS (OTTO VON BISMARCK'S PROP) IN THE BATH. NEXT, YOU VISUALIZE BRYAN ADAMS IN THE LIVING ROOM AT UNCLE JOHN'S HOUSE, IMPROVISING WILDLY ON A SAXOPHONE (ORNETTE COLEMAN'S PROP).

OVER AT GREAT-AUNT HEATHER'S APARTMENT, YOU SEE AGATHA CHRISTIE, INTERVIEWING A MONKEY (THE ORGAN GRINDER'S PROP) IN THE STUDY. MEANWHILE IN THE MASTER BEDROOM AT COUSIN FREDDIE'S, ANNETTE BENING

IS DRESSING UP AS OLIVER HARDY, STICKING ON A FALSE MOUSTACHE AND WEARING A BOWLER HAT.

MOVING ON TO COUSIN DAMIAN'S HOUSE WE FIND OLIVER NORTH, WEARING BATTLE FATIGUES AND A COWBOY HAT, SHOOTING AT GLASSES WITH A GUN (ANNIE OAKLEY'S PROP) IN THE DINING ROOM. AND FINALLY, OVER AT COUSIN LAURA'S, YOU ENVISAGE BILL CLINTON OUTSIDE IN THE YARD, PLAYING TENNIS (ANDRE AGASSI'S ACTION) AGAINST THE GARAGE WALL. >

When you have memorized all the important birthdays and anniversaries, mentally replay the scenes you've created at least five times to ensure that they are firmly consolidated in your memory.

It is a good idea to think up your own list of personalities to use in the DOMINIC system. This might sound daunting, but you don't have to list all 100 people at once – you can think up, say, 10 personalities a day for 10 days. Start by working your way from 00 to 99, converting the numbers into pairs of initials. Next, jot down any personalities you immediately associate with any of them. Then, you can return to 00 and think of characters to match all the remaining initials. Remember, too, that the people don't have to be famous – you >

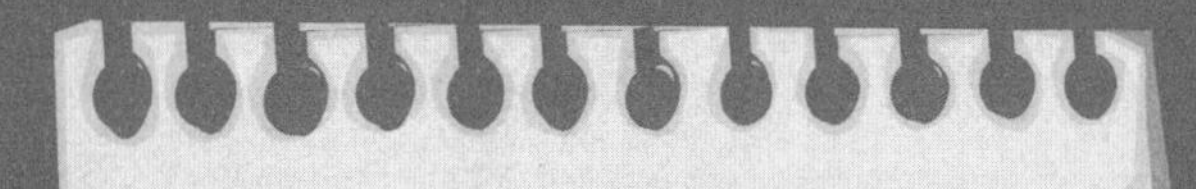

20	BO	?
21	BA	BRYAN ADAMS
22	BB	BRIGITTE BARDOT
23	BC	BILL CLINTON
24	BD	?
25	BE	BETTY EDEN (AUNT)
26	BS	?
27	BG	BARRY GIBB
28	BH	BOB HAYES (OFFICE)
29	BN	?

can include family, friends and colleagues, as well as actors, TV celebrities, politicians, singers, comedians, sportsmen and -women, and so on. They can be from the past or the present; they can be real or made-up characters. The important thing is that each person has their own unique action or prop, which you instantly associate with them (and with them alone).

WHEN YOU HAVE THOUGHT UP PERSONALITIES FOR THE PAIRS OF NUMBERS 00 TO 29, TRY THE FOLLOWING EXERCISE. READ THROUGH THE ROWS OF 2-DIGIT NUMBERS BELOW FROM LEFT TO RIGHT AS IF YOU WERE READING A BOOK, AND CONVERT EACH NUMBER INTO THE INITIALS, AND THEN THE PERSONALITY YOU HAVE CHOSEN TO REPRESENT IT.

02	12	25	17	19
14	00	08	22	29
03	07	18	21	01
16	20	26	10	11
13	04	05	24	15
06	28	09	27	23

ONCE YOU HAVE FOUND PERSONALITIES FOR ALL 100 2-DIGIT NUMBERS, YOU CAN PRACTISE THEM IN THE

SAME WAY, BY EXTENDING THE EXERCISE TO INCLUDE NUMBERS 30 TO 99.

FIRST, TAKE A PIECE OF LINED PAPER AND DRAW FIVE COLUMNS. NEXT, TAKE ANOTHER PIECE OF PAPER, WRITE OUT THE NUMBERS 00 TO 99 AND THEN CUT THEM OUT. PUT ALL THE CUT-OUT NUMBERS IN A BOWL, JUMBLE THEM UP, AND, AS IF YOU WERE HOLDING A RAFFLE, DRAW OUT A NUMBER. NOW WRITE THIS NUMBER IN ONE OF THE COLUMNS. CONTINUE DRAWING NUMBERS AND WRITING THEM IN ROWS FROM LEFT TO RIGHT ACROSS THE COLUMNS UNTIL YOU HAVE USED ALL THE NUMBERS AND HAVE 20 ROWS OF 2-DIGIT NUMBERS ACROSS THE FIVE COLUMNS. YOU CAN NOW READ THE NUMBERS IN EACH ROW, FROM LEFT TO RIGHT, CHANGING THEM FIRST INTO INITIALS AND THEN PERSONALITIES. IF YOU FEEL CONFIDENT, YOU CAN ADD IN THE PERSONALITIES' PROPS OR ACTIONS, TOO.

THE GOLDEN RULE OF MEMORY

Try to repeat and recall any memorizations at least five times to ensure that they are firmly embedded in your memory. For example, to ensure that you memorize a bank account number, retrace the steps of your journey and go over the associations you created five times after you first made the memorization.

TIME TO REMEMBER

Our ability to remember is influenced by many factors, such as how tired we are, whether we have just eaten, even what time of day it is. Research has shown that, for most of us, the best time to perform memory tasks is during mid- to late morning, and the worst time is for about an hour immediately after lunch.

If you think you could be a future World Memory Champion, try the following advanced test, which involves memorizing the mathematical value of pi (π) – the number that expresses the ratio of a circle's circumference to its diameter. At school you might remember learning $\pi = 3.141$, but in fact π is an irrational number – it cannot be expressed as an exact fraction or as a decimal with a finite number of decimal places. Below I have listed the first 100 decimal places of π. See how many you can memorize accurately.

3·14159265358979323846264 3
38327950288419716939937510
58209749445923078164062862
089986280348253421170679 ›

I suggest you use a combination of the Journey Method and the DOMINIC System. First, think up two 25-stage journeys. Next, divide the numbers into pairs of digits, so that you can turn them into personalities and their props and actions. Then, place four numbers: two personalities (or one person and the other's prop or action) at each stage. To recall the numbers simply retrace your journeys and write down each pair of numbers.

How did you do? The current world record for the memorization of pi stands at 42,195 decimal places (established by Hiroyuki Goto of Tokyo, Japan in 1995), so you probably have a bit of work to do before you'll be able to break this. But keep practising, and maybe I'll see you at the next World Memory Championships!

1st

ACKNOWLEDGMENTS

AUTHOR'S ACKNOWLEDGMENTS

I wish to thank the creative team at Duncan Baird Publishers, including Judy Barratt, Bob Saxton and Dan Sturges, for producing this book, and especially Ingrid Court-Jones for her invaluable work.

Managing Editor: Judy Barratt
Editor: Ingrid Court-Jones
Designer: Dan Sturges
Commissioned Artwork: Maggie Tingle